# VOLCANIC
# NEW ZEALAND

# VOLCANIC NEW ZEALAND

Photographs by Eric Taylor Text by James Cole

Best wishes
Jim Cole

Auckland
Oxford University Press
Melbourne Oxford

*Oxford University Press*
Oxford London Glasgow
New York Toronto Melbourne Auckland
Kuala Lumpur Singapore Hong Kong Tokyo
Delhi Bombay Calcutta Madras Karachi
Nairobi Dar Es Salaam Cape Town
and associates in
Beirut Berlin Ibadan Mexico City Nicosia

First published 1983



ISBN 0 19 558101 6

Designed by Ross Ritchie
Photoset in Century Expanded by Monoset, Auckland
Printed by Hip Shing Printers Ltd, Hong Kong
Published by Oxford University Press,
Trentham House, 28 Wakefield Street, Auckland, New Zealand

# Contents

# Acknowledgements

I would very much like to thank the individuals and companies who have helped with the production of this book. They are: the staff of Heli Tours (Rotorua) for their very generous assistance, Taupo Air Services Ltd., Volcanic Wunder Flites Ltd., Kodak New Zealand Ltd., Professor R.H. Clark, Dr Jim Cole, Mr Greg Gosson, Dr Mary Livingston, Mr Brian Brake, Mr Brian Campbell (Senior Wildlife Officer, Turangi), Mr Peter Ferris, Mr Phillip Patterson, Ms Wendy Harrex, Bubbles H. Mihinui (Guides Supervisor, Whakarewarewa Thermal Reserve), Tim and Terry of Orakeikorako, Myrie Randell and my mother, Mrs E.G. Taylor.

The photographs were taken with Nikon cameras and lenses, using Kodachrome film.

ERIC TAYLOR

# Introduction

According to Maori legend, it was Ngatoro-i-rangi, the famous navigator and tohunga of the Arawa canoe, who created volcanic activity in the Central North Island of New Zealand. The Arawa canoe landed at Maketu in the Bay of Plenty and its people spread inland to settle in the Rotorua and Taupo areas. Ngatoro is said to have left his home at Maketu and travelled south until he reached the great plain to the east of Taupo where he and his companions rested. While there, the clouds lifted and Ngatoro saw Mount Tongariro. He was amazed at its beauty. He declared the mountain tapu and decided he must climb to the top. He called to his companions and they continued their journey southwards until they reached the lower slopes of the mountain.

'Stay here,' Ngatoro commanded his companions. 'I will climb to the top of the mountain, but none of you may come with me except for the slave Auruhoe. While I am gone you must eat no food. This will give me strength and the gods will be with me. When I return we will feast together and I will tell you the things that I learn from the mountain.'

So Ngatoro and Auruhoe set off. For a start the walk was pleasant, through bush and tussock, but as they climbed it became colder until they were struggling through snow-covered slopes near the top of the mountain. Cold numbed their faces, hands and feet, and Auruhoe fell but Ngatoro urged him on.

The journey took some time and Ngatoro's companions became hungry. 'Perhaps they have perished, and we will wait for ever. Let us eat,' they said and broke the fast.

The gods were angry at this disobedience and sent blizzards to the mountain so that it became even colder and Ngatoro and Auruhoe became frozen. In his agony, Ngatoro prayed to his sisters in Hawaiki, his legendary homeland. 'Send fire to warm me or I shall perish,' he called.

In Hawaiki his sisters heard the prayer and called the fire gods Te Pupu and Te Hoata. They plunged into the sea and swam quickly until they reached Whakaari (White Island) where they lifted their heads, leaving fires which today form the White Island volcano. They found they still had some way to go and so dived below the surface again rising occasionally at Motuhora (Whale Island), Okakaru, Rotoehu, Rotoiti, Rotorua, Tarawera, Orakeikorako and Taupo, to make sure they were still on course. Finally, they burst through at Tongariro. The fires warmed Ngatoro and he gained strength, but Auruhoe succumbed to the intense cold.

Ngatoro gave praise to the fire gods and, as a gesture of his gratitude, picked up the body of Auruhoe and cast it into the newly formed crater. Ever since, the cone has been known as Ngauruhoe.

In geological terms, volcanoes occur where hot *magma* from inside the Earth is able to escape along fractures or pipes and reach the surface. The Earth is very hot inside but has a cool rigid crust which is a mosaic of separate moving *plates*. Under this rigid crust is the more mobile *mantle*, which is slowly circulating due to convection currents. Where these currents arise, cracks develop in the rigid crust and volcanoes form.

In the Pacific Ocean this is happening along a line extending from Mexico to an area south of New Zealand — the East Pacific Rise and Pacific-Antarctic Ridge (see diagram).

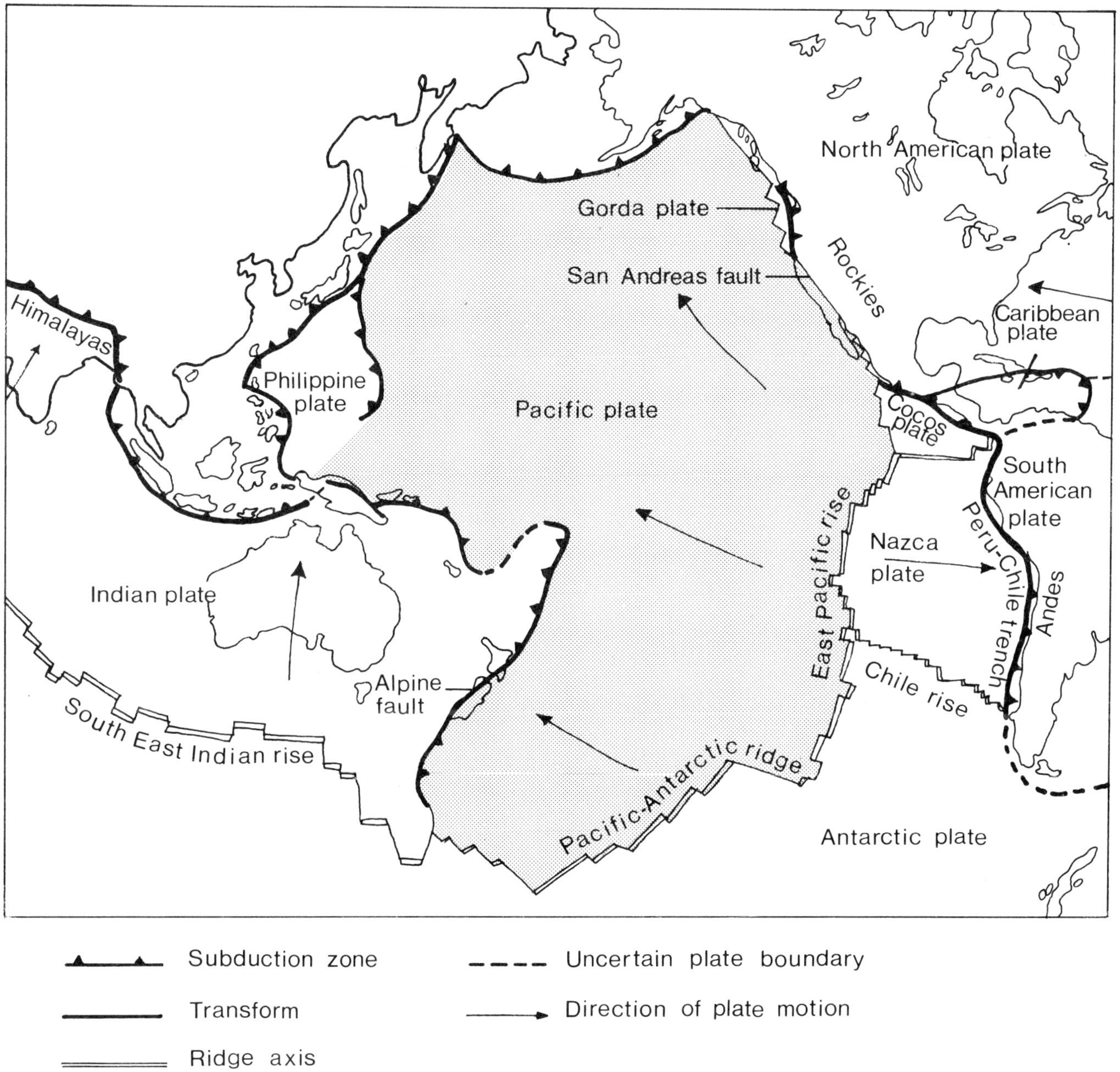

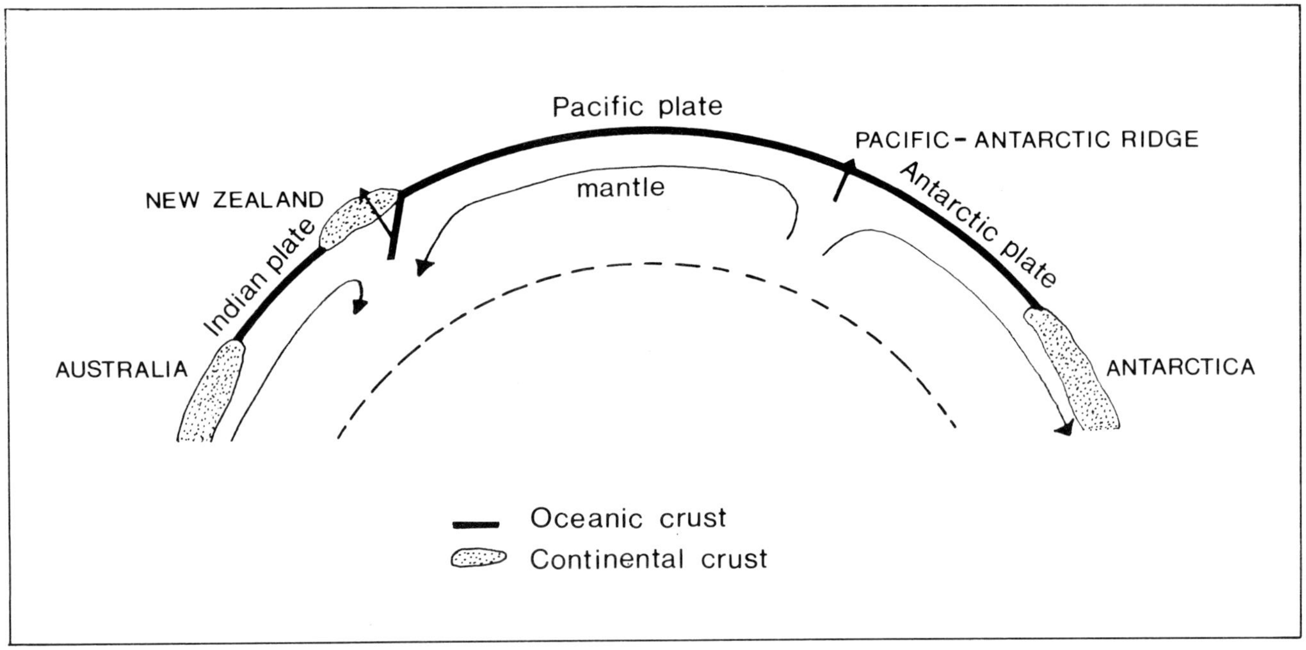


Cross section through the S.W. Pacific region.

The Pacific crustal plate is subducted beneath New Zealand, generating magma which rises to the surface to form andesite volcanoes.

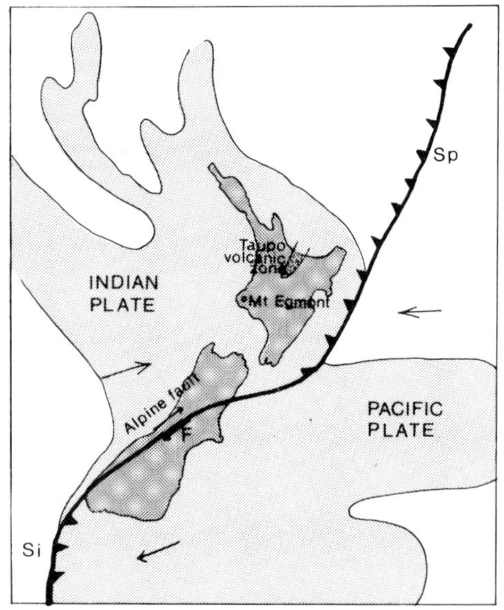


←Direction of movement of the plates
Sp Pacific plate being subducted beneath the Indian plate
Si Indian plate being subducted beneath the Pacific plate
F Two plates sliding beside one another
Continental crust

Plate boundary across New Zealand.

Geologically, this feature is termed a *mid-ocean ridge* and is a place where new oceanic crust is forming. It is a continuous process and, as the volcanic rock cools, it pushes the previously formed oceanic crust along like a conveyor belt on the mobile mantle below. Eventually, this 'conveyor belt' will meet another rigid plate and one of two things may happen. It may slide past the plate, with the junction marked by a major geological fault line such as the famous Alpine Fault of the South Island of New Zealand. Such features are the location of numerous and often very large earthquakes. Alternatively, one plate may slide beneath the other plate. This forms a *subduction zone*. As the plate descends into the mantle it will heat up and partially melt to produce a different kind of magma which will rise to the surface to create a second line of volcanoes.

Volcanoes of the North Island of New Zealand are related to a subduction zone in which the Pacific plate on the eastern side descends beneath the Indian plate to the west. The resultant line of volcanoes extends from White Island in the north to Ohakune in the south and is usually referred to as the *Taupo Volcanic Zone*.

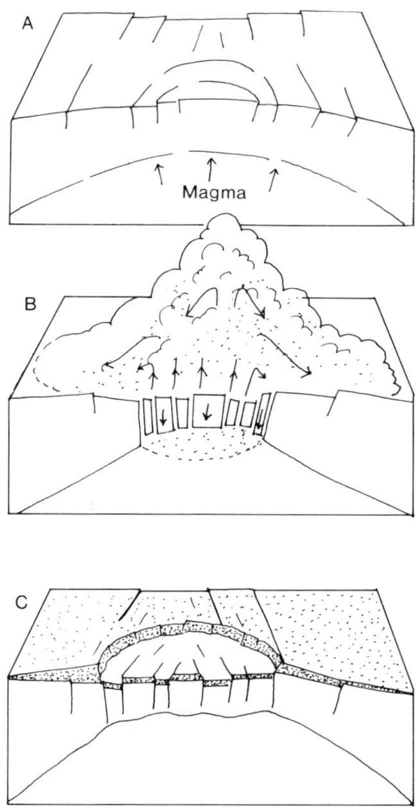


Sequence of events during a typical ignimbrite eruption.

A – Pressure causing build-up of gas pressure in magma.
B – Large ignimbrite eruption.
C – Collapse to form caldera.

The Taupo Volcanic Zone is fascinating not only from a geological viewpoint but also scenically because of the wide variety of features present. Most major types of volcano occur, from the classic volcanic cone of Ngauruhoe to Lake Rotorua, which is in geological terms a *caldera* but which topographically gives little indication of the violence of its past volcanic history.

Broadly speaking, we can divide the volcanoes of the North Island according to their composition. The large volcanoes of the Tongariro National Park are made of *andesite*. This type of lava is erupted fairly quietly to form lava flows alternating with explosive eruptions which may form volcanic ash. These volcanic materials build up cones like Ngauruhoe, usually with a surrounding apron of ash. Usually there are several centres of eruption which overlap to produce a larger volcanic structure like Tongariro or Ruapehu.

Another type of volcanic material which is common in the Taupo and Rotorua areas is made of *rhyolite*. This type of lava contains more silica — the main constituent of bottle glass — and is more sticky than andesite. As a result it does not form lava flows like andesite but rather forms large mounds or domes above the vent through which it is erupted. These domes may develop to considerable size and create imposing structures like those forming Mt Tarawera.

Perhaps the most extraordinary type of volcano and volcanic rock is the *ignimbrite*. During eruption it is the most devastating phenomenon imaginable — there have been many such eruptions during the history of the Taupo-Rotorua region. Fortunately, no one has ever seen a large ignimbrite eruption anywhere in the world but by studying the deposits left by them we can reconstruct what must have happened.

When rhyolite magma is highly charged with gas it may break through to the surface with a violent explosive eruption. Sometimes the gas-charged ash will blast to a height of many kilometres and spread out over a wide area. The ash will then rain down and collect as layers on the ground surface. At other times the magma will boil over the surface, rather like milk boiling over a saucepan, and the gas-charged mass will sweep over the countryside often for hundreds of kilometres, destroying everything in its path and leaving behind a flat-topped deposit that may be a hundred metres thick in places. This is an ignimbrite and the deposits of such eruptions from the Taupo-Rotorua area have formed the flat-lying Kaingaroa and Mamaku plateaux on either side of the volcanic zone.

Volumes of these deposits may be enormous. Some are over 100 cubic kilometres. Such volumes of magma erupted from beneath a volcano must create an unstable situation, and in order to take up the space left by the erupting magma the overlying rock caves in to leave a large crater or caldera. This the origin of Lake Rotorua and of the northern part of Lake Taupo.

Eruptions are periodic events but after an eruption some magma remains in the crust beneath the volcano. This will take a long time to cool but while doing so provides a heat source within the region. Rain water seeping through the rocks becomes heated and starts to circulate by convection, rather like water in a beaker over a bunsen-burner flame. The hot water will rise within the convection system and some of it will reach the surface to form geysers and hot springs, also typical of the Rotorua-Taupo area.

The water is likely to be slightly acid and so will tend to break down some of the volcanic rocks near the surface. The ultimate result of this is the formation of fine clays which form the basic ingredient of the mud-pools, another feature of the area.

The volcanic area has considerable interest for the historian. Maori history and legend are necessarily interwoven but there seems little doubt that many of the original canoes that came to New Zealand (the Arawa has already been mentioned) landed in the Bay of Plenty and parties moved inland to settle in the Rotorua and Taupo areas. The story of Ngatoro-i-rangi (who is credited with bringing volcanic activity to the region) has already been told, but there are many others. For example, Rotorua is said to have been claimed for the Arawa tribe by Ihenga, grandson of Tama-te-Kapua, chief of the Arawa canoe. When Ihenga arrived at Rotorua he saw smoke and realised that other people had settled their first and had already set up a tuahu or sacred shrine by the lake. The posts of the tuahu were old and partly burnt, indicating they had been there for some time. There was nobody in the village when Ihenga arrived so he conceived a plan to claim the area for the tribe. He exchanged the old posts of the existing tuahu for a new one that he had brought with him. He then set up the old ones at his own camp some distance away. As the occupants of the settlement were returning, Ihenga met the chief and claimed the area as his own. The chief naturally challenged him but Ihenga asked what proof the chief had that he was there first. 'That is easy,' the chief responded. 'The tuahu will surely establish our rights to this land.'

'Very well,' said Ihenga. 'By the tuahu shall proof be made. The older tuahu shall have right of occupation.'

When they entered the village, the chief knew he had been tricked but he bowed to the cunning of Ihenga and Ihenga took possession of Rotorua for the Arawa tribe, who lived there from that time on. They used hot pools for bathing, boiling pools for cooking, and the hot ground for heating their houses.

Much later a number of tribal battles took place in the volcanic region. One of the most famous of these was between the Arawa and a war party of the Ngapuhi tribe from the Bay of Islands, led by the famed Hongi Hika.

The Arawa were expecting the attack, as it was in revenge for an earlier attack by the Arawa which killed a number of Ngapuhi. The Arawa had gathered for safety on Mokoia Island in the centre of Lake Rotorua, taking all the available canoes with them. Hongi Hika and his group carried their canoes inland from Tauranga making their way over lakes and through bush to Lake Rotoiti; one of the routes between Lakes Rotoehu and Rotoiti still bears the name 'Hongi's Track'. For several days Hongi sailed round Mokoia Island before finally attacking. According to Arawa tradition this attack took place on a misty morning and the Arawa sentries did not see the approaching Ngapuhi — it was left to a flock of tarapunga (gulls) to sound the alarm. Throughout the battle the tarapunga circled, watching the conquest of the Arawa, and it is said that the spirits of the Arawa dead entered the bodies of the gulls which are to this day tapu. The Ngapuhi won the battle and withdrew, taking many prisoners with them.

Battles continued throughout the Land Wars, but eventually peace came and the attractions of local thermal activity began to draw increasing numbers of tourists. The main settlement was at that time the Maori village of Ohinemutu and as this expanded Rotorua was established as a 'spa' town. Today it is a thriving tourist resort and together with Taupo forms the main business centre for the volcanic region.

Tourism is very important to the volcanic area, but it is not only the thermal attractions that people come to see. The volcanoes of Tongariro National Park are snow-covered during the winter months and Ruapehu becomes a mecca for skiiers. On a fine winter weekend thousands of people flock from the main centres to ski down its slopes.

The rivers and lakes of the region are famous for trout and anglers come from all over the world to try their luck. Many of the pools of the Tongariro and other rivers draining into Lake Taupo are ideal for trout and many weigh over three kilograms. Trout were first introduced into New Zealand in 1868 when brown trout eggs were imported from Tasmania; rainbow trout eggs, imported from California, followed in 1883. A feature of the rainbow trout in New Zealand is that careful conservation has kept the strain pure, whereas in its native California interbreeding has meant that this is no longer the case there.

Industry has also come to the volcanic area. The ignimbrite plateaux make an ideal site for the growth of exotic forest. The climate and soil allow species such as *Pinus radiata* to reach maturity in only 25-30 years, and commercial production can start as little as ten years after planting. The Kaingaroa forest on the eastern side of the volcanic region is the largest man-made forest in New Zealand and one of the largest in the world. More than 111,000 hectares of exotic forest have been planted, of which 72 per cent is in radiata pine, the most economic tree to grow. The operation involves nearly 2,000 kilometres of road and 400 kilometres of firebreaks, a nursery of over 10 million trees and a workforce of 1,500. Current wood production is nearly 30 million cubic metres annually.

Some of this timber goes to one of a number of pulp and paper mills sited in the volcanic region. One of these is sited at Kawerau, south-west of Whakatane, where geothermal energy is used to provide the power necessary for the mill's operation as well as to supply steam for the process used.

The volcanic area has great potential. The majestic volcanoes, bubbling hot pools and clean, clear streams and lakes should ensure that the area is popular for both New Zealand and overseas holidaymakers. So it is essential that the natural resources in the region are well-managed and conservation is not only preached but also practised. It is in the national interest that some of the enormous energy stored in the volcanic area be used for power generation, but some of the thermal areas must be left in their natural state if the geysers and hot springs are to be preserved.

It is also important that we remember that the area is volcanically active. Eruptions will take place in the future and the more we learn about past eruptions the more likely we are to be able to predict future events.

For the present, the area is to be enjoyed and the photographs Eric Taylor has collected show some of the wonders of the region. Those who know the area will recognise many of the features; one would hope that those who do not will be inspired to visit this spectacular area.

# White Island

White Island is an active volcano which lies 50km offshore in the Bay of Plenty. Its Maori name is Whakaari, which means 'to make visible' and like many other areas in the volcanic region it figures prominently in Maori legend. The story of Ngatoro has already been told, but another legend links Whakaari with Motuhora (Whale Island). It is said that in times when mountains walked, talked, loved and fought, Whakaari and Motuhora lived in the Huiarau Range north of Lake Waikaremoana. One night the two mountains decided they wanted to visit the sea and during the night rushed down to the coast. On their way Whakaari gouged out the Whakatane river valley and Motuhora, who was tall and slim, the narrow gorge of the Waimana River. When they arrived, Whakaari sent her sister to find food, but while she was gone, dawn began to break. Whakaari knew that as soon as the sun's rays touched them, they would become fixed to the spot, so to see as much as possible she raced out to sea. When Motuhora returned she saw her sister and chased after her, but the sun's rays caught them before Motuhora had caught up with Whakaari, which explains why Motuhora is sited closer to the shore.

The facts are much less romantic. Both islands are part of the chain of volcanoes extending from White Island to Ruapehu. Whale Island is the older of the two and is now extinct, but White Island remains very active.

White Island was named by Lieutenant (later Captain) James Cook who, after rounding East Cape on 31 October 1769 in HMS *Endeavour*, wrote in his journal: 'The land seen yesterday bearing west and which we now saw was an island bore SW distant 8 leagues. I have named it White Island because as such it always appear'd to us.'

It is a small island, 2.7 kilometres long and 1.5 kilometres wide; the highest peak, Mt. Gisborne (321m), lies just west of the island's central point. On the southeastern side of the island is a large crater whose floor is rarely more than 30 metres above sea level. There are numerous craters and *fumaroles* in the floor of the crater, some discharging strongly acid water which forms a small stream and finally flows into the sea. The fumaroles also produce a great deal of steam, equally acid, which stings the face and is most unpleasant to breathe. Wind eddies cause the steam clouds to change direction without warning, making the island very uncomfortable at times.

There have probably been many eruptions from White Island in the past, building up layers of lava and ash. At some stage in the island's history there must have been a lateral blast (rather like the one which occurred at Mt. St. Helens in the United States in 1981). This left the present crater where all subsequent activity has taken place. For most of the nineteenth century, there was a lake in the crater but in 1913 this was drained by a company extracting sulphur.

Sulphur was first mined from White Island in 1885, but the terrible conditions forced the operation to be abandoned in 1900. Another company started in 1911, but in 1914 there was a landslide from the crater's backwall which covered the crater floor and killed all 11 people working on the island. Ten years later yet another company tried to make a success of it. They built a new factory and continued for some years, but the economic conditions of the 1930s finally beat them and operations ceased in 1936. The remains of the factory can still be seen on the island: the concrete and metal have been badly corroded but some of the timber is still in reasonable condition.

Nowadays the only regular visits are undertaken by scientists who periodically monitor the volcano to check on its eruptive activity. Since 1967 they have found that the crater floor swells slightly before an explosive eruption and then deflates once the eruption has taken place. This was observed before ash eruptions in 1968 and 1971, and as a result an eruption on 26 December 1976 was able to be predicted. This eruption was the first to erupt new lava since observations of the island began. The composition of the lava proved to be very similar to that erupted from Tongariro and Ngauruhoe, emphasising the link between the volcanoes of the chain.

This eruption has created a new sub-crater in the floor of the main crater. This sub-crater continued to enlarge until in 1980 the floor was well below sea level.

The island is a truly alien place. There is no fresh water in the crater, and little sign of life. A few scrubby plants occur away from the fumaroles and occasionally the tracks of a kiore (Maori rat) can be seen. Outside the crater, pohutukawas grow in isolated patches and a gannet colony lives for part of the year on the southern slopes. The sea around the island is, however, very rich in marine life. Hot springs discharging offshore seem to attract fish and hence it is a popular fishing ground.

From offshore, the island can appear fascinating and worth a visit. It is privately owned, however, and permission should be sought before landing. For those intrepid enough to attempt it strong boots are essential and a gas mask is desirable. It is certainly no place to go if ill-prepared!

White Island (Whakaari), from the south. White Island is an active volcano situated in the Bay of Plenty, 50km northeast of Whakatane. The island is 2.7km long and 1.5km wide, and since 1976 has been extremely active.

An aerial view of the crater of White Island. The mounds on the crater floor were formed from a flow of volcanic mud which swept down in 1914.

A fumarole vent on Donald Mound, White Island. The gas emitted from the crater is about 600°C, and the extreme heat has caused the sulphur to burn, giving it a grey appearance. At night the vents on Donald Mound glow red.

White Island. A view looking north from South Crater Bench towards the 1978 crater, showing patterns made by rain water in a recent ash fall. Some idea of scale is given by the tiny human figures in the bottom right of the picture.

A view towards the bottom of the 1978 crater, White Island. At 150 metres in depth, this is the deepest and most active crater on the island. The most recent lava eruption on the island occurred here in 1977.

# Rotorua

The volcanic area, with its many unusual and often spectacular features, is sometimes called the "tourist diamond of New Zealand" and Rotorua lies at its heart. The city is a marvellous centre to explore the area despite 'Soir de Rotorua' — the ever present odour of hydrogen sulphide which pervades the city. This distinctively volcanic smell offends many people when they first arrive but it is quite remarkable how short a time it takes to get used to it.

Rotorua sits at the southern edge of one of two adjacent rhyolite volcanic centres. In the distant past these centres were the sites of great ignimbrite eruptions which caused the land to collapse and form calderas. The Rotorua volcanic centre is a single caldera, 16km across, most of which is filled by Lake Rotorua. Some time after the gigantic eruption that formed the caldera, lava oozed out quietly at several places to form lava domes. The most obvious of these form Mt Ngongotaha on the western side of the lake and Mokoia Island in the centre.

Because of the prominence of these features it is natural that they have figured in Maori legend. Mokoia Island, for example, is the setting of perhaps the greatest Maori love story — the romance of Hinemoa and the young chief Tutanekai. Tutanekai lived on Mokoia Island and once, on a visit to the villages on the shores of Lake Rotorua, he met Hinemoa, and the two fell in love. Hinemoa's family strongly disapproved and the two were forbidden to see each other. Tutanekai returned to Mokoia Island but the two could not bear to be parted. Before he left, the two lovers developed a plan. Hinemoa would creep out at night, take a small canoe and paddle across the lake guided by the sound of Tutanekai's flute. On the first night the music of the flute came soft and clear across the water and Hinemoa crept down to the water, but to her dismay the canoes had all been dragged up the beach and into the bush and she did not have the strength to carry them into the water. So she was forced to creep back into bed. The next night she tried again but still the canoes were beached. Night after night she tried to no avail, until she could stand it no longer. Secretly, she emptied six gourds, cast off her cloak and tied the gourds firmly around her body. She slipped quietly into the water and swam across to the island. At first the sound of the flute was loud and clear but after a while the music stopped, for Tutanekai had given up hope. Eventually, she reached the island, cold and tired. As she staggered up the beach she came upon a hot pool and lay in it soaking up the luxurious warmth. Next morning, she saw Tutanekai's slave getting water for his master. Disguising her voice she asked for a drink and then deliberately smashed the gourd. The slave ran to Tutanekai and told him what had happened. Tutanekai was furious, ran to the pool and issued a challenge. Hinemoa stepped from the pool and the two were united.

Okataina volcanic centre lies to the east of Lake Rotorua. It too was the site of large ignimbrite eruptions, but here there have been many subsequent eruptions — some quiet, forming rhyolite domes, others explosive, covering the area in a thick mantle of volcanic ash. As a result the area has a wild hummocky topography with numerous picturesque lakes filling the depressions between the lava domes.

One of the most beautiful of these lakes is Lake Okataina. There is no road around the lake and the only way to explore it properly is by launch. Along most of the shore native bush, dominated by giant tree-ferns, reaches to the water's edge, but occasionally there is a small secluded beach on which to relax.

Many of the lakes are well-known to fishermen as they contain rainbow and brown trout. For the non-sportsman, however, the best places to see the trout are the streams on the western side of Lake Rotorua. These have been developed as tourist parks and visitors can watch large numbers of trout, some of which have grown to enormous size, at very close quarters. There are specially constructed viewing windows along some of the streams giving a fascinating underwater view of the trout.

Most of the central part of the Okataina centre is relatively inaccessible. It is covered by forest and bush and provides some of the timber needed by the big Kawerau pulp and paper mills. On the southern side of the centre is Tarawera volcano. This is surely one of the most spectacular scenic features in New Zealand. The rhyolite domes which form the main part of the volcano were rent open by an eruption on 10 June 1886, which exposed the inner parts of the volcano. This has provided a unique opportunity for scientists to study the inside of a volcano as well as providing a colourful and unforgettable sight for tourists.

The rhyolite domes have built up over 20,000 years. We know eruptions occurred about 19,000 years ago, then 15,000 years ago, 11,000 years ago and finally about 600 years ago. The domes grow internally (rather like blowing up a balloon) until finally they resemble upturned pudding basins. Often these domes will act as a plug, trapping gas in the underlying magma. Pressure builds up and explosive eruptions of ash and pumice occur.

The eruption which took place 600 years ago formed the present skyline of Wahanga, Ruawahia and Tarawera peaks. The volcano then lay dormant. It was regarded as tapu by the Maoris and few Europeans ventured beyond the lower slopes. The pink and white terraces, two separate and enormous fan-like silica terraces on the shores of Lake Rotomahana, directly south-west of Tarawera, were well-known, however, and drew visitors from all over the world. To serve them a settlement was established on the shore of Lake Tarawera at Te Wairoa and a two-storeyed hotel was built.

On 31 May 1886 two separate tourist parties on their way across Lake Tarawera to view the pink and white terraces reported seeing a fully dressed and manned war canoe speed across the lake. No such canoe existed, and the event caused widespread excitement. The presence of this 'phantom' canoe was taken by local Maoris as an omen of impending disaster, but Europeans viewed the claims with predictable scepticism.

However, on 10 June 1886, the Maoris' concern was proved to be justified. At 12.30 a.m. a series of earthquakes began, increasing in intensity until 1.30 a.m. when a relatively small explosion occurred on the north-eastern side of Mt. Tarawera. At 1.45 p.m. there was a roar from near Ruawahia peak and a black column of *basaltic* ash glowing with reflections from red-hot rocks rose above the volcano.

Another explosion occurred at 2.10 a.m. from the southwestern end of Tarawera. The whole mountain was now in eruption. At 3.30 a.m. a series of violent explosions occurred from a series of craters across the mountain and at the same time there was a violent eruption from the Rotomahana basin and a column of steam and mud rose even higher than that from Tarawera. The mud which fell from the cloud was plastered over the surrounding countryside, several metres of it falling on the village of Te Wairoa, causing the hotel to collapse.

The eruption ceased at 5.30 a.m. as rapidly as it had begun. At least 116 people were dead, a chain of nine craters had formed across Tarawera, the pink and white terraces were destroyed, and basaltic ash and mud had been spread over most of the Bay of Plenty and parts of Hawkes Bay. It had been a very different type of eruption from the previous ones at Tarawera and, in volcanological terms, it had been a small event, but the results were still devastating.

The mountain is now quiet and visitors can view the results of the eruption. In many of the craters three distinctly coloured units can be seen. Vertical walls with a pink or grey hue are the 600-year-old rhyolite domes. Above these is white banded pumice from an explosive part of the same eruption. Then at the top of the section is the red and black basalt erupted in 1886; the red colouration is due to oxidation of iron during cooling of the lava. Surrounding the craters are many large rocks scattered over the surface. These were thrown out during the final violent explosive phases of the eruption.

Apart from small wisps of steam in some of the craters on Mt. Tarawera there is no thermal activity remaining on the mountain. To the southwest, however, is Waimangu, a spectacular thermal valley that did not exist before the eruption of 10 June 1886. Until then a small stream had flowed down the Haumi valley into Lake Rotomahana, but during the eruption a series of explosions left ten craters along the length of the valley. Afterwards, those nearest Lake Rotomahana were covered by a rise in water level of the lake, but those at the south-western end remain as a tourist attraction. Of these, the two most spectacular are Ruaumoko's Throat and Echo Crater.

Ruaumoko's Throat (sometimes called Inferno Crater) is a vertical-walled explosion crater which contains a sky-blue lake. The level of this lake fluctuates up to ten metres: at about six-weekly intervals it overflows and a river of hot water flows down into Frying Pan stream below. After a few days, the lake level recedes and the river dries up.

Echo Crater is a large basin filled by Frying Pan Lake. This hot-water lake, which constantly bubbles due to the release of gas and steam, covers an area of over 4.85 hectares and is reputed to be the world's largest 'boiling' lake. Periodically, explosions occur within Echo Crater. The most spectacular of these was Waimangu Geyser which broke out from the north-eastern end of Echo Crater in late 1900. This geyser continued intermittently until 1904 and commonly threw muddy water and blocks up to 150 metres and occasionally as high as 488 metres above the crater. It was unpredictable — on 30 August 1903 four people were killed during one of its violent outbursts.

Another geyser developed to the southwest of the site of Waimangu geyser in 1917. It began without warning, directing a steam-blast to the south-west which lifted the roof off the accommodation house nearly a kilometre away. The steam badly scalded the occupants, two of whom later died. The geyser lasted only a few days and then stopped. After this, the area remained quiet until 1973 when another small *hydrothermal* eruption took place. This one only lasted 15 minutes and caused no damage.

Perhaps the most famous of the thermal areas in the Rotorua area is Whakarewarewa on the southern side of Rotorua city. The area contains many hydrothermal features in a zone one kilometre long and half a kilometre wide on either side of the Puarenga stream.

Two types of hot spring can be recognised at Whakarewarewa. These are called chloride springs and sulphate springs according to their dominant chemical composition. Chloride springs are clear and discharge hot water very near boiling point. This water usually contains high concentrations of silica ($SiO_2$), derived from the underlying volcanic rocks, and chloride salts, such as sodium chloride (common salt). When the water reaches the surface small amounts of the silica dissolved in the water may come out of solution and be deposited. Over the course of time this will form *sinter* terraces, good examples of which can be seen at Whakarewarewa.

Sulphate springs are highly acidic and are generally cloudy or form mud pools. The acid water and steam alter the rocks to form clays, particularly *kaolin*, and it is this which gives the mud pools their consistency. Iron or other metals are usually present and these create the colours frequently associated with the pools.

Probably the most popular of the thermal attractions are the geysers, and at Whakarewarewa Pohutu is the most consistently active. It erupts frequently, usually to a height of about 18 metres, although this will depend on local water conditions. All of the thermal attractions of the area are very susceptible to hot water being drawn off in the surrounding region. As a result, thermal bores sited in the southern part of Rotorua city must be closely watched. If too much water is drawn off, Whakarewarewa could dry up altogether.

The reason for geyser activity is complex, but in simple terms it depends on water being able to 'dissolve' more gas or steam when under high pressure than when under low pressure. The same principle applies to a bottle of champagne. When the stopper is released from the champagne bottle, pressure is reduced and the dissolved carbon dioxide is released, sometimes explosively!

Under the geyser there are commonly many chambers filled with hot water. In deeper chambers the water is under pressure, but when the hot water rises towards the surface, pressure is reduced and the gas and steam comes out of solution to form a mass of bubbles. These bubbles increase the volume and a hydrothermal eruption occurs. Once the upper chambers are emptied the eruption will cease and the geysers will remain quiet until pressure builds up again.

Rotorua is a fascinating place. There are so many things to see, many of which can be seen in few other places around the world, and at the end of the day what can be nicer than to relax in one of the many hot thermal pools available within the city and contemplate the next day's outing?

Whakarewarewa. Pohutu geyser at sunset, with the Prince of Wales geyser to the right. Pohutu erupts to a height of 18 metres.

Champagne Pool, the largest hot pool in the Waiotapu thermal field. It covers an area of 7000 sq. metres, and has a surface temperature of around 75°C.

An aerial view of Ruaumoko's Throat, overflowing into Frying Pan stream below. Mt. Tarawera and Lake Rotomahana are visible in the background.

Close-up beside the Champagne Pool. The orange precipitate coating the siliceous sinter is rich in metals in particular arsenic, antimony, gold, silver, mercury and thallium.

Warbrick Terrace, a sinter apron which has formed in the Waimangu Valley since the Tarawera eruption. The colours of the terrace are largely due to the growth of algae on the surface.

Dawn at Hell's Gate, Tikitere.

An aerial view of Mt. Tarawera, under spring snow. Tarawera last erupted on 10 June 1886. Red and black scoria was spewed from a fissure running across the mountain.

Champagne Pool. Maungaongaonga, a dacite volcano can be seen behind the lake, and two small areas of steam can be seen rising from a fumarolic area on the southeast flanks of the volcano.

Sinter pattern beside the Maggie Papakura geyser, Whakarewarewa.

Mt. Tarawera. A view north along the 1886 rift, from near the Ruawahia Trig. The higher area on the left of the photograph is Wahanga Dome formed during the Kaharoa eruption about 600 years ago. In the distance the Edgecumbe, Whale Island and White Island volcanoes can be seen forming a line extending into the Bay of Plenty.

Explosion debris, Hell's Gate, Tikitere.

Iodine Pool, Waimangu Valley.

Echo Crater at dawn. This hot-water lake covers nearly five hectares and is reputed to be the world's largest boiling lake. Cathedral Rocks are seen in the background.

Whakarewarewa. Mahanga or Boxing Glove geyser, which erupts almost continuously, to a height of 3 metres.

Mt. Tarawera. During the latter stages of the eruption the eruptive vents blocked, pressure built up and there were violent explosions creating the series of craters shown in the photographs.

Te Ara Mokoroa Pool, Waimangu Valley.

A view across Geyser Flat, Whakarewarewa, at sunset.

# Taupo

The full name for Taupo is Taupo-nui-a-Tia (the great shoulder mat of Tia). The reason for the name is somewhat obscure, but Tia was a member of the great Arawa canoe and he is thought to have spent some time at Taupo.

The town is located at the northern end of Lake Taupo and the approach from the north on a good day can be spectacular. There is a broad panorama with Taupo in the foreground, the expanse of Lake Taupo behind and in the background the peaks of Tongariro, Ngauruhoe and Ruapehu.

Early Maori settlers in the area were descended from people who arrived in the Arawa canoe. They formed the Tuwharetoa tribe who came from the Bay of Plenty, probably in the

1700s, allied with a smaller group of Ngati Kurapoto, and took possession of the area. When the first European missionaries arrived in about 1839 the area had quite a large Maori population. The first European settlement was built in 1869 when an Armed Constabulary post was established during the campaign against Te Kooti — a reminder of this post can be found in the naming of the present-day A.C. (Armed Constabulary) baths in Taupo. Thermal attractions in the area soon made the town popular but its comparative isolation meant that it was not until the 1950s that Taupo really became a holiday centre. Today it is a major tourist centre with excellent hotels, motels and facilities to cater for everyone.

Much of Taupo's appeal is obviously centred on the lake. Lake Taupo is the largest lake in New Zealand (619 square kilometres) and is a popular venue for yachtsmen and fishermen alike.

According to Maori legend the lake formed when Ngatoro-i-rangi, en route to the Tongariro volcanoes, climbed to the summit of nearby Mt. Tauhara and hurled a tree from the top. The tree pierced the ground below, causing water to well up and Lake Taupo was formed. In fact, the lake owes its origin to some very large volcanic eruptions.

A particularly violent eruption occurred about a third of a million years ago. The first stage of the eruption may have been small, spreading a thin covering of ash over the surrounding district, but after a while a great surge of hot pumice ash and rock debris swept over the countryside, first filling valleys but then spreading over some 2500 square kilometres to form a massive ignimbrite. About 150 cubic kilometres of material was erupted and the source area collapsed to form a caldera which is now filled by Lake Taupo's Western Bay.

About 20,000 years ago another ignimbrite eruption occurred, this time from the north-east part of Lake Taupo. This was not quite as violent as the earlier eruption but spread pumice over a wide area to the north.

In about AD 130 (less than 2000 years ago) yet another violent eruption occurred, centred on Horomatangi reef in the eastern part of the lake. This eruption also started as a small explosion, but after a while the explosions became more and more violent, sending gas, ash and pumice into an enormous cloud which may have reached 40 kilometres or more into the air. The volume of solid material within the cloud was so great that it collapsed, sending a massive flow of pumice and ash over the surrounding region. Much of this filled river valleys and some was carried out to sea, later to be washed up on the beaches around the North Island. The finer ash was carried by wind to eventually fall as a thin 'tephra' (a Greek word for ash) over much of the North Island.

Today, Lake Taupo looks peaceful but there is no reason to suppose its history of violent volcanic activity is over. Fortunately, such eruptions are very infrequent and dormant periods are measurable in tens of thousands of years. We can only hope the next eruption is a very long way off.

Ten kilometres north of Taupo township there is another expression of the Earth's potential volcanic energy: the Wairakei geothermal scheme, which was brought into operation in 1958. High pressure water and steam from underground reservoirs is tapped by a series of boreholes and the steam is piped to a power station built on the banks of the

Waikato River. Until recently, the hot water was wasted. It was separated from the steam and, still under pressure, passed to concrete silencing towers where pressure was suddenly released, causing the water to 'flash' into great columns of steam. Now some of this water is flashed in controlled stainless steel towers and the steam is recovered and piped to the power station.

In 1979, the station had a load capacity of 192 megawatts, sufficient to provide about five per cent of the country's total power consumption. So it is a very valuable resource, but its development has not been without cost. Wairakei valley used to contain a number of spectacular geysers. In the 1930s Wairakei Geyser, for example, sent up a column of water nearly 20 metres high every ten minutes, but no longer does so. The sinter terraces remain, as do the many areas of steaming ground and some hot-water basins, but the tourist park is only a shadow of its former grandeur.

Further down the Waikato River, another tourist attraction has been changed by Man's insatiable need for energy. Orakeikorako was well-known to the Maoris as a thermal area long before European settlement, but it was not until 1937 that walking tracks were established and it became a popular tourist park. In 1961, the area underwent a major change in character. Filling of Lake Ohakuri behind the newly constructed Ohakuri Dam caused the water level in the Waikato River to rise about 20 metres at Orakeikorako, flooding the lower hot springs and geyser and the site of the tearooms, which had to be relocated.

The thermal area is now reached by jet boat across Lake Ohakuri. The remaining attractions have if anything been enhanced by the rise in water levels of the lake below. Diamond Geyser plays intermittently to a height of about three metres; behind it is a series of terraces of which the Golden Fleece Terrace is perhaps the most spectacular. This is a mass of white sinter nearly five metres high, cascading over an old fault line from the Artist's Palette above. Perhaps the best known feature at Orakeikorako, however, is Ruatapu (sacred cave) or Aladdin's Cave. The entrance is framed by tall tree-ferns and in the floor of the cave is Waiwhakaata, a hot pool which was used by Maori women long before Europeans visited the area.

Taupo, like Rotorua, is a good place to explore the features of the volcanic area. A good general view of Lake Taupo and the surrounding area can be obtained from the top of Mt. Tauhara, behind Taupo township. This is an old volcanic lava dome, whose name means 'lonely mountain' (this may refer to its isolated position or to its unsuccessful love affair with Pihanga in the Maori legend described later in this book). Afterwards, a drive around the lake can be well worthwhile. Many of the streams are popular fishing spots and at certain times of the day there are so many fishermen trying their luck that they form a 'picket fence' across the river mouth.

At the southern end of the lake are the settlements of Tokaanu and Waihi, which have developed around another thermal area. Waihi is a fascinating place. It has twice suffered from landslides which have swept down the hills behind the village. On the first occasion, in 1846, a landslide swept through Te Rapa pa (400 metres from the site of present-day Waihi) killing all but three of the inhabitants, including the Tuwharetoa paramount chief, Te Heuheu Tukino II. His son was absent from the pa at the time and earned the name

Horonuku (landslide). Horonuku later gave much of Tongariro National Park to the nation. A second slide came down in 1910 but fortunately this time only one person was killed. The village has several buildings which are magnificently carved in the true Maori style.

The return drive to Taupo can be made on the western side of the lake, where the pink-coloured ignimbrite erupted from Western Bay can be seen in roadside sections. A short detour can be made to Kinloch, one of many lakeside settlements and a popular base for the many boats used on the lake, before returning to Taupo township.

Waihora Bay, Lake Taupo, at dawn.

Afternoon storm, Motuoapa Peninsula, Lake Taupo.

Mt. Tauhara at sunset. Tauhara is an old volcanic lava dome, 1099 metres high, whose name means 'lonely mountain'.

Cascade Terrace, Orakeikorako. The terrace was formed by the deposition of silica from hot water erupted from the Cascade geyser above it. The colours are largely due to algae.

Golden Fleece Terrace, Orakeikorako, eight metres high and 40 metres long. A geyser which first erupted in 1982 is visible on the right.

Diamond Geyser, Orakeikorako.

Extinct geyser, Orakeikorako.

Palette Pool on the Artist's Palette, Orakeikorako. The Artist's Palette is a sinter terrace bounded on the northwest side by the Golden Fleece terrace. Activity varies so that at times springs discharge boiling water while at other times it is dry with water found only in basins immediately surrounding the springs.

Emerald Terrace, Orakeikorako. The photograph shows rivers of hot water with deposition of sinter. Between the rivers, different algae give different colours. The area shown here is about 20 by 30 metres.

Ruatapu Cave, Orakeikorako.

Rainbow Terrace, Orakeikorako. This is a sinter terrace cascading over a scarp where the Rainbow Fault crosses the valley of the Waipapa Creek. The colouration is largely due to algae, although some of the brown colours may be due to clays stabilized by the algae.

Geothermal development at Wairakei. When bores tap the underground hydrothermal reservoir a mixture of steam and hot water rises to the surface. The steam is under pressure and can be piped directly to the turbines. Some of the water is piped to 'flashing units' at the well head where pressure is released suddenly in a closed chamber. If the resultant steam has sufficient pressure it too is sent to the turbines but through a different pipe. The remainder of the water, and steam under lower pressure goes to a concrete silencer where it is allowed to discharge as the familiar clouds of steam.

# Tongariro

The volcanoes of Tongariro National Park are perhaps the best known of the volcanic region. Activity began in the area several hundred thousand years ago, since when the products of numerous eruptions have constructed the large andesite volcanic *massifs* of Ruapehu, Tongariro, Pihanga and Kakaramea, together with a number of smaller centres at Ohakune, Hauhungatahi and Pukeonake. Ngauruhoe, which is commonly regarded as a separate volcano, is geologically a young cone of the Tongariro massif.

Tongariro was the first national park created in New Zealand. It was donated by chief Te Heuheu Tukino IV (Horonuku) who in 1887 gave the mountain to the government 'for the purposes of a National Park' because he feared the consequences of European development of the area. The mountains had for a long time been a special place to the Ngati Tuwharetoa tribe, who buried their chiefs on the mountainside and had designated it tapu.

The government accepted Horonuku's offer and in 1894 an Act of Parliament formally constituted the Tongariro National Park. At that time the park comprised only that land within the radius of a mile of the volcanic peaks and totalled 2630 hectares. Progressively more land was added until the park now covers over 75,000 hectares.

During its early years of operation the park was fairly inaccessible and visitors were few, but with completion of the main North Island railway line between Wellington and Auckland in 1908 its popularity increased dramatically. In 1919 it was decided to develop the western slopes of Ruapehu into a winter sports resort, and the Bruce Road was constructed, further improving accessibility.

Today the national park is one of the most popular in New Zealand. In summer, many people make use of the well-maintained tracks and excellent huts to explore the volcanoes and enjoy the magnificent scenery, while in winter Ruapehu becomes the North Island centre for winter sports. There are ski-fields with excellent facilities at Iwikau at the top of Bruce Road on the western side of the mountain, and at Turoa at the top of Ohakune mountain road on the southern side. Tukino, on the eastern side of the mountain, is currently under development and is used by a number of ski clubs from the main centres.

Ruapehu is the most southerly volcanic massif in the park and is constructed of alternating lava flows and thick layers of blocks and ash which were produced during explosive eruptions. The first eruptions of the volcano were centred to the north-west of the present-day summit, and the lavas and other erupted material have been strongly eroded by water and ice so that only remnants of this original phase of activity remain. They tend to form jagged ridges such as Pinnacle Ridge, on the north-eastern side of Whakapapa Valley.

Later activity was centred on the present summit area. At least two craters can be recognised, with the currently active vent containing a crater lake. This lake has existed for at least a hundred years, as its presence was recorded during a climb by George Beetham in 1879. Periodically, there are small explosive eruptions from the lake and occasionally a larger event occurs. In March 1945, a dome of lava about 40 metres in diameter developed in the lake. This soon collapsed but in May was replaced by a larger dome which displaced the water, causing it to flow out through an ice cave and into the Whangaehu River on the eastern side of the mountain. Activity continued to increase until on 22 August there was a violent explosion sending ash over the surrounding countryside. Similar explosions occurred throughout September and October but by the end of the year activity had ceased. By January 1946 water started to fill the crater again and within a few years the lake had reached its old level.

Before the eruption, excess water draining into the lake flowed out through the ice cave and into the Whangaehu River so that the lake level remained nearly constant. During the eruption, however, ash and debris blocked the ice cave to form a 'plug' so that water could no longer escape. As a consequence, water level continued to rise until by August 1953 it was eight metres higher than before the eruption. This extra water created a lot of pressure on the ash and debris in the cave, and just before 10 p.m. on December 24 1953 the plug was pushed out and the lake level dropped six metres in two and a half hours. The water rushed down the Wangaehu River, picking up boulders and other debris, to form a *lahar*. This reached the railway bridge at Tangiwai at 10.15 p.m., sweeping away one of the piers moments before the Wellington to Auckland express train was to cross it. The train plunged into the river, killing 151 people.

Since this time smaller explosive eruptions have occurred on a number of occasions from the crater lake. These are particularly dangerous if they occur in the winter as hot ash and mud is thrown out which melts the surrounding snow to form lahars that rush down the nearest valley. One of the most susceptible is the Whakapapa valley, which contains the Iwikau skifield. Fortunately, the main eruptions during the last 20 years have been either outside the skiing season or at night, but it remains a great danger and recently a lahar warning scheme has been installed in the valley. In event of a lahar skiers will at least have a few minutes' warning, which should give them time to get out of the valley bottom before the lahar reaches them. Close surveillance of the volcano is obviously essential.

Tongariro lies to the north of Ruapehu and has perhaps a greater variety of volcanic features. It is probably of similar age to Ruapehu, but has a larger number of individual volcanic craters and cones within the massif, many of which have formed in the last 20,000 years. Explosive eruptions scattered ash around the surrounding district and where this killed vegetation the charred logs have been radio-carbon dated to establish the date of the eruption. One such eruption took place about 9700 years ago and by careful study of the thickness and distribution of the ash, it can be shown that the ash was probably erupted from the crater on the north-eastern side of Tongariro that is now occupied by the freshwater Blue Lake. Similar explosions occurred from the southern part of Tongariro to form six craters, two of which now form the picturesque Tama Lakes.

When Tongariro is viewed from the north, one of the most prominent features is North Crater. This is a flat-topped cone which must originally have been a crater similar to that occupied by Blue Lake. After formation of the crater, however, it was filled by lava to form a lava lake. This broke through the rim on the western side of the crater to form a short lava flow. As the lava in the crater cooled and solidified, it may have trapped some gas in the volcanic vent for later there was a small explosion from the north-western part of the crater. This has left a pit 300 metres in diameter and 60 metres deep in the smooth top surface of the congealed lava lake.

One of the most spectacular features on Tongariro is Red Crater. This is a young cone composed of red and black scoria which contrasts markedly with the light grey lavas of the older Tongariro volcanic centres. There have been a number of eruptions from Red Crater, during one of which lava rose as a vertical sheet or *dike* and poured out onto the surface as a lava flow. Once the eruption ceased, lava in the top of the dike drained back into the volcano, leaving an unusual vertical cavern in the crater wall. Lava flows from Red Crater can be seen in Oturere Valley to the east, Central Crater to the north and South Crater to the west. The most recent eruptions have been small explosions which were recorded in 1855 and 1926, but the crater floor is still steaming, and some of the steam and gas has discharged into the Emerald Lakes just north of Red Crater. The gases have condensed in the cold water, in which sulphur and other minerals form a fine suspension, giving the lakes a milky appearance.

The most active centre on Tongariro is Ngauruhoe. This cone was first formed about 2500 years ago and probably reached most of its height in a relatively short space of time. Lava flows down the side of the cone alternated with explosive eruptions scattering ash over a wide area, killing much of the surrounding beech forest. After this initial burst of activity eruptions became more widely spaced, still alternating between lava extrusions and explosive ash eruptions.

The most recent eruption of lava occurred in 1954, when a series of flows came down the western side of the cone and into Mangatepopo Valley. Since 1954, there have only been relatively small explosive eruptions, but some of these have been spectacular. In January and March 1974 and February 1975 huge columns of ash-laden gas were erupted from the volcano, reaching up to nine kilometres above the summit. Much of this was spread over the surrounding countryside, but on several occasions the volume in the column became so great that debris swept down the north-western side of the cone as an avalanche of blocks and ash which reached Mangatepopo Valley. During the largest eruption on 19 February 1975, large blocks were strewn over the slopes of the cone. Many of these were sufficiently hot to be plastic on impact.

The scenery on Ruapehu and Tongariro is spectacular, as shown by the photographs, but it must always be remembered that they are active volcanoes and should be treated with respect. Weather can also change very rapidly on the mountains and it is always wise to be well-prepared before tramping on them. However, if the right precautions are taken, exploring the National Park can be most rewarding and thoroughly recommended.

Mt. Ngauruhoe at dawn. The cone is 2291 metres high.

Mt. Tongariro. An aerial view of Tongariro volcano from the northwest. In the foreground is North Crater with its familiar explosion pit and small flows of lava spilling over the western rim. Behind is the youngest cone on Tongariro-Ngauruhoe, with the lava flows which erupted in 1954 clearly visible on its northwestern slopes.

An aerial view of Mt. Ngauruhoe. Mt. Ruapehu rises to a height of 2797 metres in the distance.

An aerial view looking down into the crater of Mt. Ngauruhoe.

Ketetahi Hot Springs, on the slopes of Mt. Tongariro.

Close-up of sinter. Sinter is made up of a variety of forms of silica. Most is common opal but other forms such as chalcedony, jasper, agate and crystalline quartz may occur. This photograph illustrates the variety of colours possible in the sinter.

Emerald Lakes, Mt. Tongariro. The milky appearance is caused by a suspension of sulphur and other minerals in the water.

Emerald Lakes, after a storm.

Red Crater, Mt. Tongariro, composed of red and black scoria. During an eruption, lava poured out as a vertical sheet, on to the surface. After the eruption had finished, the lava drained back into the volcano, leaving a vertical cavern visible in the crater wall.

Afternoon storm, Mt. Ruapehu, seen from the Desert Road.

Aerial view of Mt. Ruapehu in midwinter.

Crater Lake, Mt. Ruapehu. Ruapehu has had a crater lake since observations first began. The water in it is highly acid and has a high percentage of dissolved solids, including sulphur which usually gives the lake a dark grey colour. Sulphur commonly floats on the surface as a black slick. The temperature varies markedly. In 1982, for example, it ranged from 13°C to 57°C. When the lake is hot, convection sets up and the sulphur slick is carried towards the sides of the lake revealing the dark grey colour of the water. Occasionally water at the bottom of the lake gets too hot and a hydrothermal eruption occurs.

# Auckland

The city of Auckland is built on many small basaltic volcanoes. Over 50 eruptive centres can be recognised ranging from cinder cones with small lava flows to explosion craters from which little or no volcanic material has been erupted. Activity from each centre was on a relatively small scale and the total volume of material erupted is less than five cubic kilometres. About half of this volume occurs at Rangitoto, whose most recent eruption may have occurred as recently as 225 years ago, making it by far the youngest volcano in the Auckland area. This volcano forms the classic low-angled cone in the Waitemata Harbour that can be seen from many vantage points around Auckland city. It is a fascinating place to visit. The jagged, clinkery lava is of the blocky or 'aa' type (the term comes from Hawaii where this type of lava is common). At the summit several small cinder cones rise to a maximum height of 259 metres above sea level.

Most of the remaining centres in the Auckland area have probably formed in the last 50,000 years, making an average of one every thousand years. Many of these centres also started their eruptive episode by forming lava flows and then, perhaps as the basalt in the vent started to solidify, became explosive to form one or more cinder cones on top of the flows. Good examples occur at One Tree Hill and Mt. Eden where the cones were used for Maori pa sites.

The explosion craters are commonly filled with water and now provide valuable recreational areas such as the Panmure and Orakei Basins.

There is no indication that volcanic activity in the Auckland area has ceased. It is dormant. Whether the next eruption will be in ten, a hundred, a thousand or ten thousand years is impossible to say, but judging from past activity, if and when it does occur, it will most likely be small and probably at a different site from the existing volcanoes. Rangitoto, however, has had more than one eruptive episode and in view of the geologically recent date of the last eruption, could also be a possibility for future activity.

An aerial view of the crater of Brown's Island (Motukorea), with Rangitoto in the distance. Rangitoto is the youngest and largest of the Auckland volcanoes, and dominates Waitemata Harbour.

Brown's Island from a height of 1000 metres. The crater and mounds visible here are the remains of a volcano created when the harbour was dry land. Most of it has been subsequently worn away by sea action.

# Taranaki

In Maori mythology mountains were particularly important and were considered to have personalities of their own — in myths they fought each other and moved about the countryside carving great valleys as they went. One of the most famous of the stories concerns the mountains of the central North Island.

It is said that in bygone days Taranaki (Mt. Egmont), Tauhara, Putauaki (Mt. Edgecumbe), Pihanga and Tongariro all resided in the centre of the North Island. All were male, except for Pihanga whose bush-covered slopes sweep gracefully up from Lake Rotoaira on one side and Lake Taupo from the other.

The Maoris of the central North Island and those of Taranaki have different versions of the story. The Tuwharetoa tribe of Taupo, for example, consider that all the mountains loved Pihanga but she loved only Tongariro. So Tongariro issued a challenge to the other suitors, defeated them and left them no choice but to depart. Taranaki went west, carving out the course of the Wanganui River as he went; Putauaki chose to travel to the warm Bay of Plenty. Tauhara held back, reluctant to leave his beloved Pihanga, until dawn came and he became fixed to his present position at the north end of Lake Taupo.

Taranaki Maoris say that Tongariro and Pihanga were married. Once, when Tongariro was away, Taranaki wooed and won the graceful Pihanga's heart. Unfortunately for Taranaki, Tongariro returned to surprise the guilty pair and, in the fight that ensued, Taranaki was driven west. The depression under Fatham's Peak is said to have been caused by a kick from Tongariro during the fight. Now Taranaki mourns his lost love and frequently weeps for her, giving the high rainfall of Taranaki, or hides himself in a cloak of mist.

The idea that mountains move is absurd. Or is it? We now know that the surface of the Earth is continually moving. Plates move against one another creating earthquakes or allowing molten magma to reach the surface to create volcanoes. The Taupo Volcanic Zone is in geological terms an *extensional marginal basin* which is opening up a few millimetres per year. It is unlikely that this opening takes place all the time, but every few decades there is an earthquake or series of earthquakes, like those experienced by Taupo in 1922, when the ground moves and open fissures form. These may be several centimetres across, cumulatively opening up part of the Taupo Volcanic Zone by perhaps half a metre. At other times volcanic dikes are intruded, such as occurred at Tarawera in 1886.

So, although mountains do not move overnight and create great river valleys, mountains may move relative to one another over the course of geologic time and often the 'zone of spreading' between them is a valley or basin like much of the Taupo Volcanic Zone.

Aerial view of Mt. Egmont (Taranaki) at sunset. The mountain rises to a height of 2518 metres.

Mt. Egmont (Taranaki), with a cloud halo.

# Glossary

*andesite* a volcanic rock which commonly forms large volcanic massifs like Tongariro and Ruapehu. Andesite volcanoes typically occur in lines above subduction zones.
*basalt* a dark coloured volcanic rock, usually fine-grained, that contains a high proportion of iron and magnesium-rich minerals. In the Taupo Volcanic Zone it is commonly erupted from fissures, as happened at Mt. Tarawera in 1886.
*caldera* a large basin-shaped volcanic depression, more or less circular in form, which is formed by collapse after a large ignimbrite eruption.
*dike* a wall-like sheet of magma which intrudes into the Earth's crust. If magma reaches the surface, the dike becomes the vent for a fissure eruption.
*fumarole* a hole or vent from which gas or steam is emitted.
*hydrothermal* an adjective applied to heated or hot water-rich solutions, or to processes in which they are involved.
*ignimbrite* a volcanic rock produced by eruption of dense clouds of incandescent volcanic material, highly charged with gas and usually of rhyolite composition. These clouds spread over a very large area destroying everything in their path.
*kaolin* a fine white clay of hydrated aluminium silicate which is the final alteration product of silica-rich volcanic rocks.
*lahar* landslide of volcanic debris lubricated by water, and the deposits that result from it.
*magma* molten rock inside the Earth which is intruded into overlying strata and eventually cools to form igneous rock. Once magma reaches the surface it loses gas to form lava or is erupted explosively to form ash or ignimbrite.
*mantle* layer of molten or semi-molten rock beneath the Earth's solid crust. It is probably the source region for basalt and andesite magma.
*marginal basin* an area of crust inland from the line of andesite volcanoes which is widening by the intrusion of basalt dikes. Taupo Volcanic Zone is an example of a marginal basin.
*mid-ocean ridge* an area of the ocean floor where basaltic magma rises to the surface to create new crust. It represents a line from which sea-floor spreading takes place.
*massif* a coalescing group of andesite cones which collectively form a larger mountain.
*plate* a rigid portion of the Earth's crust, varying in thickness from 15-150kms, which may move relative to the underlying mantle.
*rhyolite* a light-coloured volcanic rock rich in silica, which forms large mounds or domes and is widespread in the Taupo Volcanic Zone (e.g. Mt. Tarawera).
*sinter* a deposit around hot springs composed mainly of silica.
*subduction zone* the zone where one plate slides beneath another as a result of sea-floor spreading. The downgoing slab heats up and partially melts to form andesite magma.